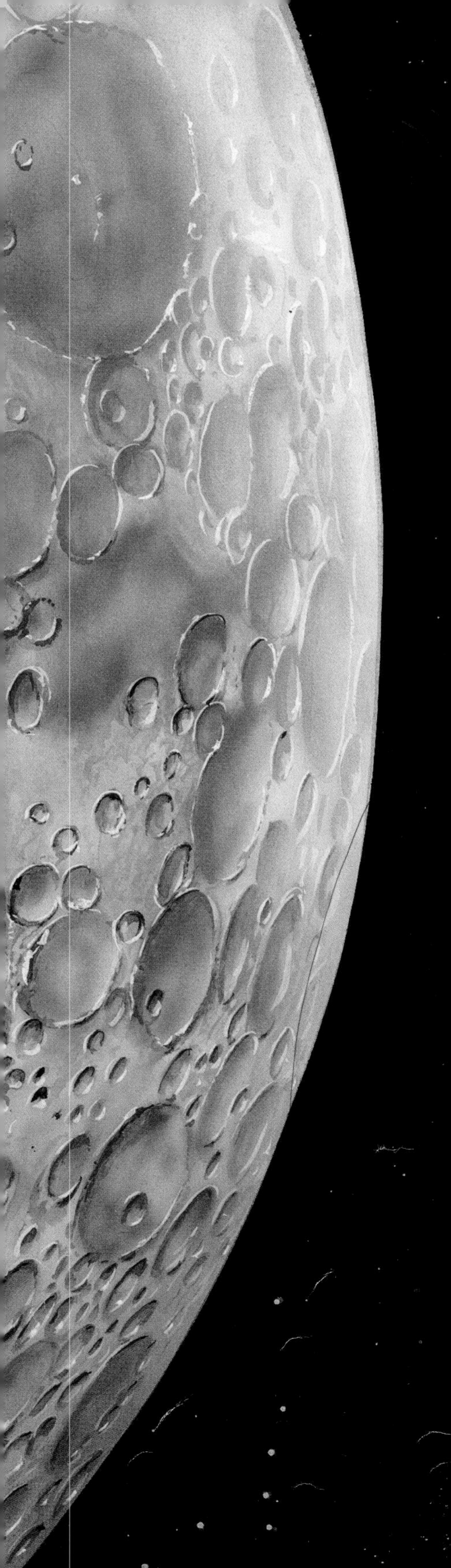

Contents

System

Earth is one of eight planets that move in space around the Sun. The Sun and all the planets make up the Solar System. The smallest planet in the Solar System is Mercury. The biggest planet is Jupiter.

Jupiter is made up of gas and liquid. It does not have a solid surface.

Jupiter

Mars

Earth

Venus

Sun

Mercury

Neptune
Saturn
Uranus
Planet of life
The planet we live on is the Earth. It is covered in water and swirling clouds. Earth is the only planet with life.
Saturn has rings around its middle. They are made of ice and pieces of rock.

The Sun

The Sun is a star in the middle of the Solar System. It is a giant ball of super-hot gases. It looks big and bright because it is a lot closer to Earth than any other star.

Sometimes, bits of the Sun explode! Hot gases shoot up from its surface into space. These are called solar flares.

Light and heat given off by the Sun keep the Earth warm and bright for us to live.

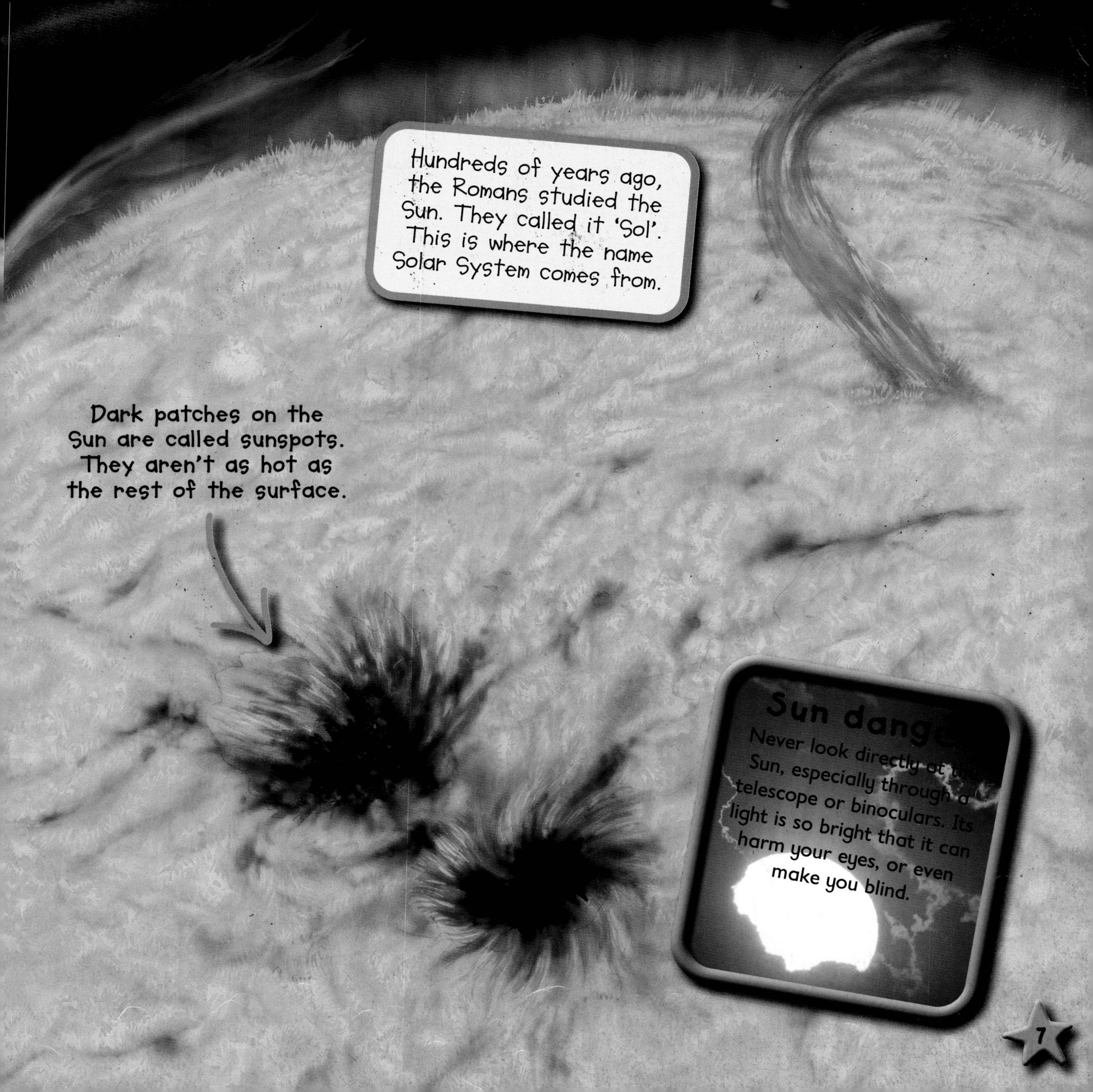

Hundreds of years ago, the Romans studied the Sun. They called it 'Sol'. This is where the name Solar System comes from.

Dark patches on the Sun are called sunspots. They aren't as hot as the rest of the surface.

Sun danger

Never look directly at the Sun, especially through a telescope or binoculars. Its light is so bright that it can harm your eyes, or even make you blind.

The Moon

Six of the planets in the Solar System have moons travelling around them. The Earth has one moon but other planets have more. These moons are smaller and stay with a planet as it travels around the Sun.

The Moon looks so bright because it reflects light from the Sun.

As the Earth spins around the Sun, the Moon spins around the Earth. It takes the Moon one month to go all the way round.

Sometimes when you look at the Moon it's possible to see strange shapes that look like a face, with eyes, nose and a mouth.

Eclipse

During an eclipse, the Moon casts a shadow that stops sunlight reaching the Earth.

The Moon is covered in dents called craters. These have been made by rocks crashing into the Moon.

Moon landing

In 1969, the USA sent three astronauts to the Moon. Two of them walked on the Moon's surface. Their names were Buzz Aldrin and Neil Armstrong. They walked around, taking photographs and collecting rocks. They even played golf and spoke to the president on the phone!

Until people landed there, no one was sure if the Moon's surface was strong enough to support people or a spacecraft.

Footprints

Astronauts have left footprints on the Moon. The footprints have not disappeared because there is no wind to blow them away.

The astronauts found that the Moon's surface was covered with fine dust and rocks.

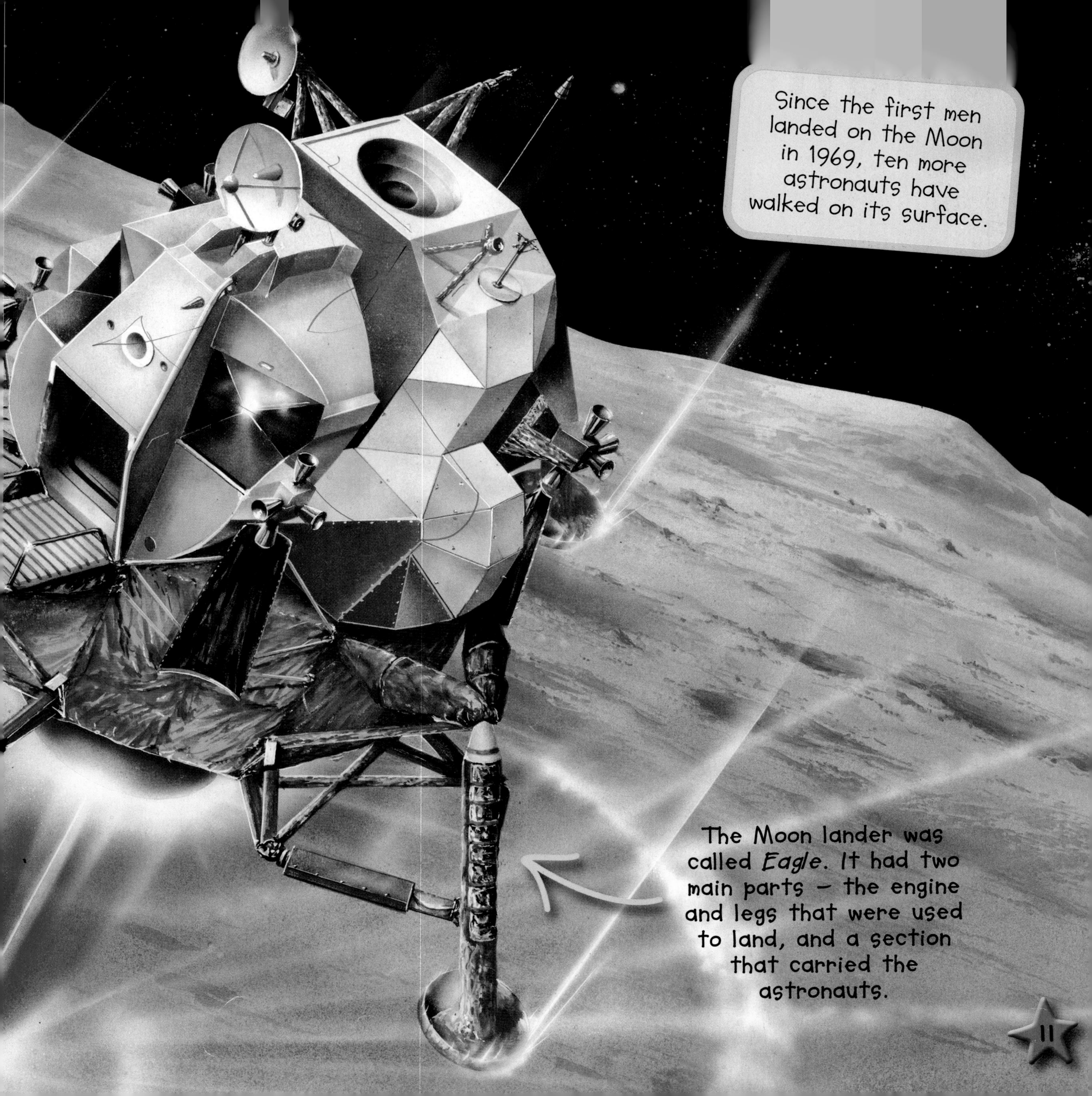

Since the first men landed on the Moon in 1969, ten more astronauts have walked on its surface.

The Moon lander was called *Eagle*. It had two main parts – the engine and legs that were used to land, and a section that carried the astronauts.

they are. They live for millions of years, and then they die. As giant stars die, they collapse and make a black hole. Smaller stars cool and fade to become a white dwarf.

As the gas and dust spin and pull together, it grows hotter and hotter at the centre. Finally, a new star begins to shine.

Stars are made out of gas and bits of dust in space. The gas and dust collect in a big cloud called a nebula.

Towards the end of its life, a star may turn into a red giant. This is very big but its surface is no longer as hot.

When the gas and dust are blown away, the new star can be seen in the night sky.

All together

Young stars stay together in groups called clusters. Eventually, they drift apart and the cluster breaks up.

The Milky Way

Our Solar System is part of a huge galaxy called the Milky Way. The galaxy is made up of around 100 billion stars. There are billions of other galaxies in the Universe. Most of them have a black hole at the centre.

Galaxy crash

If galaxies get too close, they can pull each other out of shape. These galaxies have pulled a long tail of stars from each other.

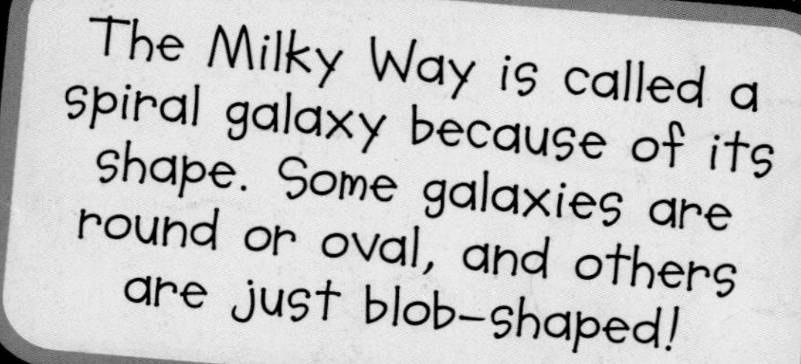

The Milky Way is called a spiral galaxy because of its shape. Some galaxies are round or oval, and others are just blob-shaped!

At the middle of the Milky Way are clouds of dust and gas. These stop scientists seeing what lies beneath them.

The arms at the edge of the Milky Way contain lots of young, hot stars that shine very brightly.

Space rocks

Between Mars and Jupiter is an enormous ring of asteroids, or space rocks. This is called the asteroid belt. Some asteroids are shaped like planets, but aren't as big. Others are huge lumps of rock, about the size of a house.

Most asteroids are about 4.5 billion years old. They are pieces of rock that were left over when the planets formed.

Jupiter is five times farther away from the Sun than the Earth is. It gets less of the Sun's heat so it would be a very cold place to visit.

At night, you can see Mars in the sky. It looks like an orange-red star. It is nicknamed 'the red planet'.

Spacecraft have been sent into the asteroid belt to look more closely at the surface of the rocks.

Meteors

If small rocks in space hit the air above Earth, they get hot and burn. We see them as glowing streaks in the sky called meteors.

Astronauts

Special suits allow astronauts to work outside their spacecraft. The suits give protection from very hot or cold conditions. They also supply air for the astronauts to breathe. Astronauts can repair spacecraft and build space stations.

Space bag

When they go to sleep, astronauts strap themselves into sleeping bags that are attached to walls. The astronauts have to go to sleep standing up!

The longest time an astronaut has spent working outside in space is nearly nine hours.

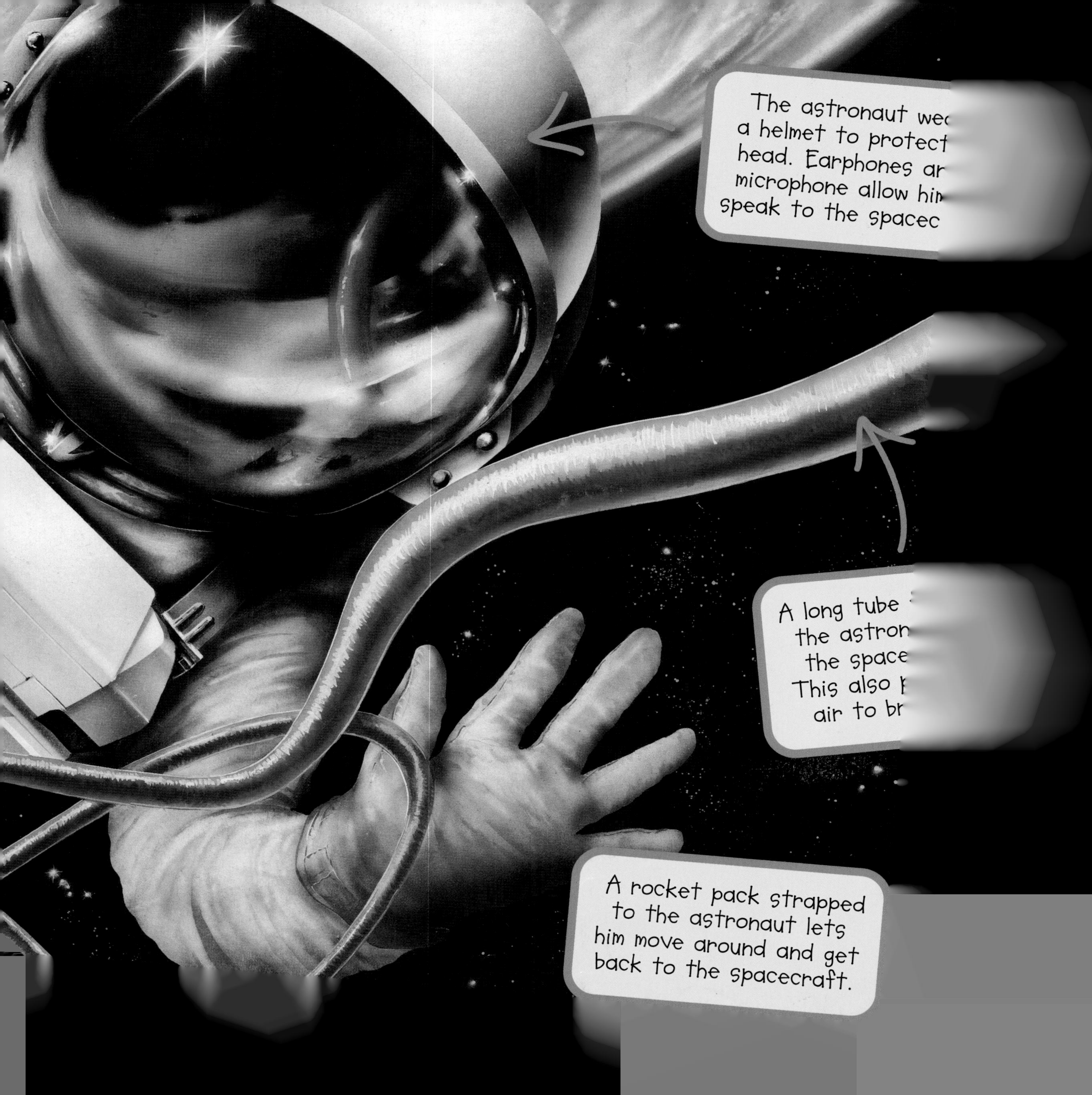
The astronaut
a helmet to
head. Earphones
microphone allow
speak to the
A long tube
the
the
This also
air to
A rocket pack strapped
to the astronaut lets
him move around and get
back to the spacecraft.

ving in space

International Space Station, or ISS
short, is still being built in space.
arate sections have been sent up by
cecraft. Once they are fixed together,
pieces make a place for
ple to live and work.

The space shuttle delivers new supplies such as food, water, fuel, spare parts and air.
Electricity used on the space station is made and stored in giant solar panels on the sides.
US

The ISS is travelling, or orbiting, around the Earth all the time. It makes just over 15 orbits every day.
Up to six astronauts can stay on the ISS at one time. They can stay in touch with their families on Earth using email and telephone.
Space plane
When the space shuttle comes back to Earth, it lands like a giant glider. It is the only reusable spacecraft.

Satellites

A satellite is a machine that travels around the Earth. Satellites gather information in space and send it back to Earth. They can take photographs of the Universe, help to tell the weather or send pictures and messages around the world.

Satellite dish

Radio telescopes collect the information that satellites send back from space.

The Hubble space telescope was launched in 1990. It helps scientists to see new and exciting things in space.

The Hubble's main camera can take detailed photos of stars and galaxies.
Solar panels use energy from the Sun to make electricity. This powers the Hubble as it travels through space.

Fun facts

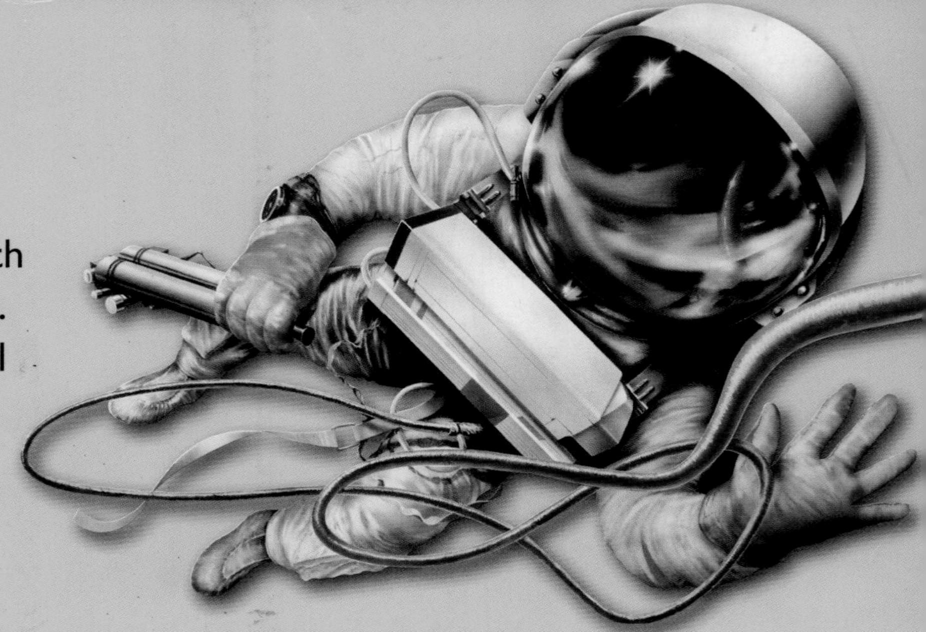

The Solar System There was a ninth planet in the Solar System called Pluto. Scientists decided that it was too small to be a normal planet, and is now classed as a dwarf planet.

The Sun The surface of the Sun is nearly 60 times hotter than boiling water. It is so hot, it would melt a spacecraft flying near it.

The Moon You would have to travel almost 400,000 kilometres to reach the Moon from Earth.

Moon landing The first astronauts on the Moon travelled about its surface in a buggy called the Lunar Rover.

Life of a star Large stars are very hot and white. Small stars are cooler and red in appearance.

Milky Way The Milky Way looks like a faint band of light in the night sky, as if someone has spilt milk across space.

Space rocks In 2001, a spacecraft managed to land on the surface of an asteroid and take pictures.

Astronauts Everything in space floats around as if it has no weight, so astronauts have special footholds to keep them still while they are working.

Living in space The International Space Station is still being built. It will be finished in 2010.

Satellites Spy satellites take pictures of secret sites around the world and listen to radio messages from ships and aircraft.